MW00510904

APPLE

Blight.—See "Pear." Remove all worthless apple trees near orchards of pear or apple.

Scab.—Spray with Bordeaux mixture (Formula 9), or lime-sulfur 1–20, just before blossoms open. Again with lime-sulfur 1–35 when petals are falling.

Mildew.—Cut out mildewed twigs as thoroughly as possible. Use lime-sulfur for scab spraying or if scab is not serious use sulfur paste, 16 pounds to 200 gallons of water (or home-made wettable sulfur spray, Formula 13 or 14) when petals are falling. Later spraying for mildew may be made with the same material. Sulfur sprays cause injury to apple trees in some sections.

Codling Moth.—Spray with lead arsenate, 2 pounds to 50 gallons of water (Formula 1 or 2) as soon as petals begin to fall and repeat in three or four weeks. In some cases it is necessary to give a third application four weeks after the second.

Red-humped Caterpillar.—See "Prune."

Tent Caterpillars, Canker Worms.—See "General Subjects."

Green and Rosy Apple Aphis.—Spray with oil emulsion (Formula 23), miscible oil, or tobacco extract (Formula 27). The critical time for application is just as the leaf buds are opening, to kill the young which are at that time hatching from the eggs.

Woolly Apple Aphis.—Spray with oil emulsion (Formula 23) or miscible oil during winter months. For root form, expose crown of roots and pour in a quantity of the above spray mixture and re-cover the roots. Nicotine sulfate (Formula 27) is also effective, or refuse tobacco may be buried in the soil over the main roots during the rainy season.

San Jose Scale.—Spray with lime-sulfur 1–10 during winter months when tree is dormant.

Tussock Moth.—Remove egg masses during winter months. Jar off caterpillars and prevent their ascent of tree by cotton or tanglefoot bands.

Flat-headed Apple Tree Borer.—Whitewash to prevent sunburn. Dig out borers. Prevent injury or wounds to the tree. The insect usually enters through dead areas.

COMBINED SPRAYING

1. For serious infestations of scale, for removal of moss, and for general clean-up, lime-sulfur 1–10 or crude oil spray (Formula 18) during the winter.

2. For green, rosy, and woolly aphis, use oil emulsion (Formula 23) or miscible oils just as leaf buds are opening. If woolly aphis is not abundant but where rosy or green aphis and scab are serious pests, use at this time lime-sulfur 1–20, to which is added one pound of 40 per cent nicotine sulfate to each 200 gallons of spray. The lime-sulfur at this time may also help to control a slight infestation of San Jose scale. Combinations of oil sprays for woolly aphis, with lime-sulfur or Bordeaux for scab, are not considered advisable. The soluble sulfur preparations (compounds with soda) may be mixed with the oil sprays and could probably be adopted for use in the control of scab, mildew and woolly aphis by determining a safe strength.

3. For codling moth and scab, use 8 pounds of basic lead arsenate in 200 gallons of 1–35 lime-sulfur, when petals are falling. For mildew add 16 pounds of sulfur paste to each 200 gallons. If rosy or green aphis appears, 1 pound of 40 per cent nicotine sulfate may also be added.

4. For codling moth and late scab infection, repeat 3, following also the recommendations for aphis and mildew if these pests need attention.

In large apple-growing regions obtain advice of local horticultural authorities as to modifications in above.

APRICOT

Black Heart.—Cut off affected branches back to main trunk as soon as the wilting is seen. Destroy badly affected young trees entirely. Avoid heavy irrigation in affected orchards.

Brown Rot, Blossom Blight, Gumming Twig Blight, Green Rot, Shot Hole.—Destroy mummies in fall. Spray with lime-sulfur 1–10 as buds are swelling and again when first blossoms are open. If rainy while "jackets" are on, spray with lime-sulfur 1–30. For the control of brown rot of the fruit, summer spraying with sulfur paste or the so-called self-boiled lime-sulfur is often recommended. In California such practice has resulted quite frequently in injury to the trees and must therefore be tried with caution. The injury consists in a yellowing of the foliage, stunting of the fruit, disfigurement by sediment on the fruit, and failure of the trees to bloom the following season.

Bud Blight.—Spray with lime-sulfur 1–8, between November 15 and December 15.

Armillaria, Crown Gall, Sour Sap.—See "General Subjects."

Black Scale, Brown Apricot Scale.—Spray during November or December with oil emulsion (Formula 18 or 23) or miscible oil spray.

Shot Hole Borers or Bark Beetles.—Remove and burn all infested portions.

California Peach-borer, Peach Twig-borer.—See "Peach."

Red-humped Caterpillar.—See "Prune."

COMBINED SPRAYING

The lime-sulfur treatment just as buds are opening as recommended for fungous diseases will also control twig-borer and give the trees a general clean-up.

ASPARAGUS

Rust.—Keep down volunteer growth. Dust tops with sulfur as soon as they are well expanded. Repeat in four weeks.

Beetle.—Cut down all seedlings except a few left for trap crop. In the spring after beetles have collected on these and deposited eggs, cut down and burn. Spray young seedlings with lead arsenate, 1 pound to 16 gallons of water, or with 40 per cent nicotine sulfate 1 gallon, whale oil soap 4 pounds, and water 20 gallons. Burn and eliminate all possible sheltering plants during winter.

Centipede or Symphyla.—Flooding in spring before cultivation has given good results in some places. Rotation of crops may be necessary for a few years.

BARLEY

(See Grain.)

BEAN

Anthracnose.—Very rare and unimportant in California.

Mildew, Rust.—Dust with sulfur at first appearance.

Wilt, Stem Rot, Rhizoctonia, Fusarium.—Prepare soil very thoroughly. Plant as late as possible, avoiding cold or wet weather. Save seed from strong, well-matured plants.

Nematode.—See "General Subjects." Black Eyes and Teparys are more resistant than other beans but sometimes badly affected.

Aphis, Thrips.—No practical field control. Nicotine sprays give temporary results if thoroughly applied, but they are usually too expensive for field work. Keep plants as vigorous as possible.

Red Spider.—If possible, keep beans thoroughly irrigated, cultivated and in good healthy conditon. Begin sulfuring as soon as spiders appear and continue throughout summer, using dry sulfur.

Weevil.—Fumigate beans in storage with carbon bisulfide 10 to 30 pounds to a thousand cubic feet of air space, the amount depending upon the tightness of the room or bin.

BEET

Curly Top (Blight).—Plant as early in season as soil can be properly prepared and good germination and growth obtained. This varies in different districts from December 1 to March 1, according to temperature and rainfall. Finish thinning before hot weather. Irrigate early, and often enough to prevent wilting during spring and summer. The disease need not be feared unless the insect *Eutettix tenella* is abundant.

Seedling Root Rot.—Replant if stand is too thin.

Rust, Leaf Spot.—No treatment needed.

Nematode.—See "General Subjets." The beet is attacked by two species, the garden nematode, *Heterodera radicicola,* which attacks a great variety of plants, and the beet nematode, *H. schachtii,* which is a pest only on the sugar beet. This does not produce large galls as with the former species.

Armyworms, Cutworms, Grasshoppers.—See "General Subjects."

Wireworms.—Plow in fall to destroy pupae. Plant early and practice clean culture. Trap adults by means of piles of straw and burn in late fall or winter.

BUSH FRUITS

(Blackberry, Loganberry, Raspberry)

Leaf Spot, Rust, Cane Blight.—Cut out and burn all affected parts in the fall. Spray with lime-sulfur 1–10 or Bordeaux mixture during the dormant season. Give good irrigation and cultivation. Renew old plantings.

Fruit Mold.—Avoid mixing moldy berries with good ones.

Borers.—Cut out dead canes and burn during winter months.

Scales.—Spray during winter with oil emulsion (Formula 23) or miscible oil.

CABBAGE

Aphis.—Spray repeatedly with oil emulsion (Formula 23) or nicotine (Formula 27).

Cabbage Worm.—Spray repeatedly up to picking time with nicotine sulfate 40 per cent, 1 pound to 100 gallons of water, or spray until two or three weeks before harvesting with lead arsenate, 2 pounds to 50 gallons of water in which 4 pounds of hard laundry soap has been dissolved.

Root Maggot.—Place tarred paper shields around bases of plants or spray repeatedly with carbolic-acid emulsion (Formula 26). Plow and work ground thoroughly in spring to destroy pupae.

CELERY

Blight.—Spray with Bordeaux (Formula 9), especially in moist weather, commencing in seed bed.

Celery Caterpillar.—Hand-pick or spray with lead arsenate.

Aphis.—Spray with nicotine (Formula 27).

CHERRY

Gummosis, Die-back.—See "General Subjects." Usually due to shallow soil or too much water. Some forms of gummosis may be parasitic. Cut out affected branches below diseased parts or cut out affected areas of bark. Sterilize as in Pear Blight. See also "Wound Treatment."

Aphis.—Spray as buds are opening with oil emulsion (Formula 23), nicotine (Formula 27), or miscible oil.

Cherry Slug.—Spray with arsenate of lead or dust with sulfur, kaolin, lime or fine road dust.

California Peach-borer.—See "Peach."

Tent Caterpillars.—See "General Subjects."

Cankerworms.—See "General Subjects."

Red-humped Caterpillar.—See "Prune."

CHRYSANTHEMUM

Rust.—Fertilize and irrigate freely to produce strong, vigorous plants.

Aphis.—Spray with nicotine (Formula 27). Cut out and burn infested plants as soon as insects appear.

Leaf Miner.—Spray with nicotine sulfate 40 per cent solution, 1 pound to 100 gallons of water.

Gall Fly.—Keep plants trimmed in spring. Spray with nicotine sulfate 40 per cent, 1 pound to 100 gallons of water when eggs appear on tips of branches in spring and summer.

CITRUS

(Grape Fruit, Lemon, Orange)

Damping-off, Armillaria Root Rot.—See "General Subjects."

Gummosis.—Cut out all affected bark as soon as discovered and treat wounds with Bordeaux paste (Formula 10). Do not allow water to stand about base of trees. In planting, keep point of budding well above ground and never allow the soil to-pile up around the trunk. For heavy soil, use trees high-budded on sour-orange root.

Scaly Bark.—Cut out all discolored bark and surface wood when the outbreak first appears. Cover wounds with Bordeaux paste

(Formula 10). Cut off badly affected branches. Watch all the trees very closely in groves where the disease is present and eradicate the disease at its first appearance.

Brown Rot, Cottony Rot.—When disease is prevalent spray lower branches and ground beneath with Bordeaux mixture (Formula 9). Do not allow boxes of fruit to stand over night in orchard. Use blue-stone in wash water (Formula 12), maintaining constant strength of 1½ pounds to 1000 gallons. Grade out very carefully all orchard-infected fruit before storing.

Blue Mold, Green Mold.—Avoid bruising in picking and handling fruit.

Mottled Leaf.—Add as much organic matter to soil as possible in form of green-manure crops, bean or barley straw, and manure. See that water penetrates to subsoil and keeps it uniformly moist. See "General Subjects."

Scales.—Fumigate with hydrocyanic-acid gas.

Mealy Bug.—Fumigate with hydrocyanic-acid gas or spray repeatedly with oil emulsion (Formula 20), crude carbolic acid emulsion (Formula 26), or water under high pressure.

Red Spider.—Dust with sulfur or spray with home-made wettable sulfur sprays (Formula 14), sulfur paste 10 pounds to 100 gallons of water, or lime-sulfur 1–50.

Aphis.—Use Nicotine Spray (Formula 27).

CORN

Smut.—Destroy diseased parts as soon as discovered. Seed treatment not effective.

Ear Mold.—Sometimes bad on late corn in moist soil. Use early varieties. Harvest and cure as early as possible. Avoid over-irrigation.

Earworm.—Clean up and burn refuse in field. Plow in fall or early spring to kill pupae. Dusting silk of ears with powdered arsenate of lead affords some relief.

Cutworms, Armyworms, Grasshoppers.—See "General Subjects."

Granary Weevil, Rice Weevil, Angoumois Grain Moth.—See "Grain."

COTTON

Corn Earworm.—See "Corn."

Red Spider.—Dust with sulfur or spray with sulfur paste or home-made wettable sulfur sprays (Formula 13 or 14).

CUCUMBER

Mildew.—Dust with sulfur at first appearance.

Nematode.—See "General Subjects."

Beetles.—Spray with lead arsenate, 2 pounds to 50 gallons of water.

Aphis.—See "Melon."

CURRANT, GOOSEBERRY

Mildew.—Dust with sulfur at first appearance.

Borers.—Cut out and burn infested twigs during winter.

Scale.—Spray with lime-sulfur 1–10, or oil emulsion (Formula 23) during the winter.

Red Spider.—Dust with sulfur or spray with sulfur paste (Formula 13), 10 pounds to 100 gallons, as spiders appear.

DAHLIA

Mildew.—See "Currant."

Aphis, Thrips, White Fly.—Spray with 40 per cent nicotine sulfate, 1 pound to 200 gallons of water.

Diabrotica.—Hand pick. Spray repeatedly with lead arsenate, 3 pounds to 50 gallons of water.

GRAIN
(Barley, Oats, Wheat)

Rust.—No remedy. Some varieties are more resistant than others.

Smut.—Seed grain should be carefully cleaned of smut balls, weed seeds and small cracked and inferior grains before treating. The smut balls in wheat and smut masses in barley may be cleaned out in fanning mills or floated out in water and skimmed. Place the cleaned seed in half-filled sacks tied at the end. Immerse these sacks for three or four minutes in a bluestone solution made by dissolving 1 pound of bluestone in 5 gallons of water (Formula 11). Drain the sacks until dripping no longer occurs, then dip them for three minutes in milk of lime made by slaking 1 pound of quicklime in 10 gallons of water. The lime prevents injury to the germ from the bluestone. If quicklime cannot be secured, air-slacked lime, 1 pound to 8 gallons of water, may be used. After thus treating, the grain should be spread out to dry, after which it may be planted or stored.

Oats are especially sensitive to bluestone and in this case it is better to use a solution of formaldehyde, 1 pound to 40 gallons of water, for ten minutes, after which no lime dip is needed. Barley is more sensitive than wheat and should always be lime-dipped after treatment with bluestone.

Armyworms, Cutworms, Grasshoppers.—See "General Subjects."

Granary Weevil, Rice Weevil, Angoumois Moth.—Fumigate in storage with carbon bisulfide, 10 to 30 pounds per thousand cubic feet of air space.

GRAPE

Mildew.—Dust thoroughly with sulfur at first appearance. Repeat if necessary.

Black Knot.—May be treated like Crown Gall (see "General Subjects") with some success. Not usually very injurious.

Little Leaf, Apoplexy, Obscure Diseases.—See "Physiological Diseases" under "General Subjects."

Leaf Hopper.—Spray with the following before insects can fly: 40 per cent nicotine sulfate 1 pound, liquid soap ⅓ gallon (hard soap 2 pounds), water 200 gallons. Clean weeds and refuse from around fences. Practice clean culture during winter.

Armyworms, Cutworms, Grasshoppers.—See "General Subjects."

California Grape Root Worm, Flea Beetles.—Cultivate thoroughly close to vines during summer and winter. As soon as bettles first appear in the spring spray with arsenate of lead, 6 pounds to 100 gallons of water.

Phylloxera.—Use resistant root-stocks.

HOLLYHOCK

Rust.—Fertilize and water freely to stimulate vigorous growth.

Caterpillar.—Spray with 40 per cent nicotine sulfate 1 pound to 100 gallons of water.

MELON

Wilt.—Plant on fresh soil. Melons cannot be grown for several years on infected ground.

Nematode.—See "General Subjects."

Aphis.—Destroy infested plants as soon as insects appear.

Flea Beetles.—Spray with Bordeaux mixture (Formula 9) as a repellant.

NURSERY STOCK

Failure to Grow after Planting in Orchard.—Very rarely due to specific disease. Usually caused by freezing, drying, or water-soaking of trees before or after planting, planting too deep, cold, wet or hot weather after planting, or some other condition unfavorable to growth. Buy of the nearest reputable nursery. Pay for good trees and see that they are handled and planted carefully. Replant all that do not grow well the first season.

Nematode, Crown Gall.—Very carefully avoid planting affected trees. The clean-appearing trees in a lot having a large percentage of infection are of doubtful value.

Deciduous.—For borers and other insects, fumigate with hydrocyanic-acid gas. Rejecting infested stock is the only safe procedure.

Citrus.—For scale insects, defoliate and fumigate with hydrocyanic-acid gas. Rejecting infested stock is the only safe procedure.

OAT
(See Grain)

OLIVE

Die-back (Exanthema).—Cut out affected branches. Add humus to soil by green-manure crops, mulch or manure. Secure uniform soil moisture and good drainage. Replace olives with plums, peaches or some other crop where die-back is very bad. See "Physiological Diseases" under "General Subjects."

Knot (Tuberculosis).—Cut out at first appearance. Disinfect as in Pear Blight.

Dry Rot, Bitter Pit.—See "Physiological Diseases" under "General Subjects." No effective treatment known except good general care.

Armillaria.—See "General Subjects."

Black Scale.—Spray with oil emulsion (Formula 22) or miscible oil during winter months.

Bark Beetle.—Cut out and burn infested branches.

ONION

Mildew.—Not successfully controlled in wet seasons. Bordeaux mixture found useful in some cases.

Thrips.—Practice clean culture. Spray with 40 per cent nicotine sulfate, 1 pound to 200 gallons of water.

Maggots.—Leave no decayed onions in field during winter. Cultivate thoroughly.

Armyworms, Cutworms.—See "General Subjects."

PEA

Mildew, Blight.—Dust with sulfur at first appearance. Repeat if necessary.

Aphis.—Repeated applications of 40 per cent nicotine sulfate (Formula 27). Field control is difficult and usually too expensive. Tobacco dust may be tried.

Weevil.—See "Bean."

Cutworms, Armyworms.—See "General Subjects."

PEACH

Armillaria, Crown Gall, Nematode.—See "General Subjects."

Little Leaf.—See "Physiological Diseases" under "General Subjects."

Blight, Curl Leaf.—Spray with Bordeaux (Formula 9), or lime-sulfur 1–10, between November 15 and December 15. Repeat with lime-sulfur when buds first start to swell.

Brown Rot.—See "Apricot."

California Peach-borer.—Dig out borers thoroughly in the fall and apply a thick coating of hot Grade C or D hard asphaltum.

Peach Twig-borer.—Spray with lime-sulfur 1–10 in spring just as buds are swelling.

Pear Thrips.—See "Pear."

San Jose Scale.—See "Apple."

Tent Caterpillars.—See "General Subjects."

Cankerworms.—See "General Subjects."

Red Spider.—See "Almond."

Combined Spraying

Two applications of lime-sulfur as recommended above will control all the usual diseases and pests of the peach tree in California which can be reached by any spray treatment.

PEAR

Black Leaf.—See "Sour Sap" under "General Subjects."

Blight.—Cut out all affected parts very thoroughly. Work especially on "hold-over" in large limbs, trunks and roots during the winter. Disinfect freely with corrosive sublimate 1–1000. Keep off all suckers and spurs from root and body. In new plantings in blight regions, top-work Bartlett and other susceptible varieties upon Surprise or other fairly immune varieties on Japanese root.

Scab.—Spray with lime-sulfur 1–12 or Bordeaux (Formula 9) just as leaf buds are opening. Repeat when first blossoms are open.

Codling Moth.—See "Apple."

Slug.—See "Cherry."

Pear Thrips.—Spray as soon as insects appear with oil emulsion (Formula 24) or miscible oil, to which may be added 1 part of 40 per cent nicotine sulfate to every 2000 parts of the spray mixture.

Blister Mite.—Spray with lime-sulfur, 1–10, as cluster buds are opening.

Red-humped Caterpillar.—See "Prune."

Root Aphis.—Use Japanese root. Eliminate young, stunted trees and replant with healthy ones. Expose roots and pour in quantity of oil emulsion or miscible oil spray (Formula 23).

Green Apple Aphis.—See "Apple."

COMBINED SPRAYING

1. When scale of any kind is abundant and for moss and general clean-up, winter spray with lime-sulfur 1–10, crude oil emulsion (Formula 18), or miscible oil.

2. For scab and thrip use Bordeaux mixture (Formula 9) or lime-sulfur 1–10 as cluster buds are opening, adding an extra 10 pounds of lime and 1 pound of 40 per cent nicotine sulfate to each 200 gallons of spray. Oil sprays cannot be mixed with lime-sulfur or Bordeaux for this purpose.

3. For scab and thrip, repeat 2 when first blossoms are about to open.

4. For codling moth and late scab infection, spray when petals are falling with 8 pounds of lead arsenate in 200 gallons of 1–30 lime-sulfur or 200 gallons of Bordeaux mixture.

PLUM, PRUNE

Armillaria, Crown Gall, Sour Sap.—See "General Subjects."

Brown Rot.—Not often serious. See "Apricot."

Tent Caterpillars, Cankerworms.—See "General Subjects."

Red Spider.—See "Almond."

San Jose Scale.—See "Apple."

Pear Thrips.—See "Pear."

Red-humped Caterpillar.—Hand-pick young colonies and spray trees with basic lead arsenate, 2 pounds to 50 gallons of water.

Brown Apricot Scale.—See "Apricot."

Black Scale.—See "Apricot."

California Peach-borer.—See "Peach."

Peach Twig-borer.—See "Peach."

Flat-headed Apple-tree Borer.—See "Apple."

COMBINED SPRAYING

For scale, moss, and general clean-up, spray in winter with oil emulsion (Formula 18 or 23) or miscible oil.

POTATO

Wilt, Dry Rot, Scab, Rhizoctonia.—Obtain clean seed from healthy plants. Discard any which show decided dark brown discoloration or decay at stem end to a depth of at least ¼ inch from the end. Soak the seed before cutting for 1½ hours in a solution of 1–1000 corrosive sublimate (1 ounce to 8 gallons of water), or two hours in formaldehyde, 1 pound to 30 gallons of water. Use a wooden vessel for the sublimate.

Jelly End, Soft Rot.—Avoid injuring and bruising in digging.

Nematode.—Use clean seed and avoid infested soil.

Armyworms, Cutworms, Grasshoppers.—See "General Subjects."

Flea Beetles.—Spray with Bordeaux mixture (Formula 9) as repellant.

Tobacco Worm, Tomato Worm.—Spray with arsenate of lead, 2 pounds to 50 gallons of water.

Wireworms.—Rotate crops, replant in spring or later if seed is destroyed. Practice clean culture.

Tuber Moth.—Cultivate thoroughly, hill vines, harvest early, fumigate infested tubers with carbon bisulfide, 10 to 30 pounds per thousand cubic feet of air space. Be sure to plant clean seed.

ROSE

Mildew.—Spray with lime-sulfur 1–10 before spring growth starts. Use dry sulfur, sulfur paste, or lime-sulfur 1–35 when disease first appears. Forty per cent nicotine sulfate, recommended below for aphis, may be added to this.

Aphis.—Wash frequently with water under high pressure. Spray with 40 per cent nicotine sulfate, 1 pound to 200 gallons of water.

White Rose Scale.—Spray with oil emulsion (Formula 18 or 23) or miscible oil in winter.·

San Jose Scale.—See "Apple."

Combined Spraying

For fungous diseases and aphis, 40 per cent nicotine sulfate may be added to sulfur sprays as given above.

SNAPDRAGON

Rust.—Water and fertilize freely to stimulate growth. Destroy badly affected plants and clean up thoroughly in fall. Pentstemon is a fairly good substitute for Snapdragons and does not rust.

SORGHUMS, SUDAN GRASS

Smut.—Controlled by seed treatment. See "Grain."

Aphis.—Water and cultivation to stimulate growth.

Grasshoppers, Armyworms, Cutworms.—See "General Subjects."

STRAWBERRY

Leaf Spot.—Clean up and burn leaves in late fall. Spray with Bordeaux mixture (Formula 9) if disease becomes serious.

Root and Stem Rot.—Use less water. Improve drainage. Wash out alkali in winter by flooding.

Aphis.—Spray with 40 per cent nicotine sulfate, 1 pound to 200 gallons of water as soon as insects appear.

Flea Beetle.—Spray with Bordeaux mixture (Formula 9) as repellant.

Crown Borer.—Eliminate and burn infested plants as soon as discovered.

SWEET POTATO

Wilt, Black Rot.—Get clean seed potatoes from an uninfested locality. Destroy diseased vines after digging.

Soft Rot.—Avoid bruising. Dry well before storing. For long keeping, pack in dry sand.

TOMATO

Damping-off.—See "General Subjects."

Wilt, Summer Blight.—Use plants free from damping-off. Replant if not too late. Cannot be controlled some years.

Late Blight, Late Rot.—Occurs in late fall or winter crop. Spray with Bordeaux mixture (Formula 9) immediately after rains.

Blossom End Rot.—Avoid drouth or irregular irrigation.

Nematode.—See "General Subjects."

Grasshoppers, Cutworms, Armyworms.—See "General Subjects."

Flea-beetles.—Use Bordeaux mixture (Formula 9) as repellant.

Tobacco Worm.—See "Potato."

Tomato Worm.—See "Potato."

WALNUT

Blight.—No specific remedy. Give trees best possible care. Thin out tops of old trees. Control aphis. Plant resistant varieties.

Melaxuma.—Cut out diseased bark areas and apply Bordeaux paste.

Crown Gall, Armillaria.—See "General Subjects."

Winter Killing.—Irrigate about November 1 if no good rains have fallen. Whitewash bodies in the fall. Do not irrigate after August, except as above.

Aphis.—Very thorough spraying with lime-sulfur 1–20 just before buds open is effective, but slow and expensive. The insects are easily killed with a summer spray of 40 per cent nicotine sulfate (Formula

27), but this method is also too slow for a large acreage of good-sized trees. Dusting with finely powdered tobacco is a promising method for rapid and effective work. With either spraying or dusting, control is much easier if the work is done early in the summer before the foliage becomes very dense and the aphis extremely abundant.

Yellow-necked Apple Caterpillar, Red-humped Apple Caterpillar.— Spray with lead arsenate, 2 pounds to 50 gallons of water. Hand-pick young colonies.

WHEAT

(See "Grain.")

GENERAL SUBJECTS

ANTS

For the Argentine ant, place a sponge in a fruit jar, saturate it with poisoned syrup (Formula 7 or 8), make a few nailholes in the cover and keep the jar in pantry and several others in the yard about the house. Add more poison from time to time, label carefully and keep away from children.

APHIDS (PLANT LICE)

Almost all species can be controlled by spraying with 40 per cent nicotine sulfate, 1 pound to 200 gallons of water. Weak oil emulsions are also effective. Tobacco dust has been found useful in some cases and is usually more quickly and easily applied than liquid sprays. All these materials kill by contact and so when the insects are on the under side of the leaves they must be actually hit with the spray to be killed.

ARMILLARIA ROOT ROT (OAK FUNGUS)

No specific treatment. Surgery as in Crown Gall or Pear Blight can sometimes be practiced on roots and crowns of trees not too far gone. Black walnut, French pear and fig roots are practically immune. Affected areas in orchard may be isolated by opening a trench 3 to 4 feet deep, around them. This may be immediately refilled if reopened every two years to keep roots cut off.

CANKERWORMS

Use tanglefoot bands or cotton bands during fall and spring. Spray with arsenate of lead, 2 pounds to 50 gallons of water.

CROWN GALL

Throw out all affected nursery trees. The clean trees in a lot having a large percentage affected are of doubtful value. In orchard, occasionally examine crown and main roots, especially of stunted trees. When not too far advanced, the galls may be chiseled out, sterilized with 1–1000 corrosive sublimate (Formula 28), and the wounds covered with Bordeaux paste or asphaltum. Badly affected and stunted trees should be pulled out and replanted, using fresh soil.

CUTWORMS AND ARMYWORMS

Broadcast freely poison bran mash (Formula 4 or 5) in front of invading insects or over infested plants.

DAMPING-OFF

Best controlled by skill in watering. Water plant beds only in morning and on bright days. Do not sprinkle oftener than necessary. In greenhouses or frames give plenty of ventilation. In making citrus seed beds, put an inch or two of clean sand on the top of the soil. Some forms of Damping-off may be prevented by steam-sterilizing the soil before planting or by drenching with a solution of 4 pounds of formaldehyde in 50 gallons of water, using 1 gallon of solution to every square foot. This must be done two weeks in advance so that no odor of formalin remains at planting time. Where Damping-off has started, spraying the plants and ground with Bordeaux may do some good.

FLIES

Should be prevented from breeding by keeping manure, garbage and similar refuse material covered tightly. For poisoning flies in the house use about 2 per cent formaldehyde solution exposed in saucers, adding a little sugar.

FROST

With young citrus trees in frosty localities, wrap trunks with corn or milo stalks in winter. Heap up earth around butts. Enclose tender valuable young trees with burlap covers. For bearing groves obtain detailed information about methods and appliances for smudging with oil fuel. (See also Sour Sap.)

GRASSHOPPERS

Scatter freely poison bran mash or citrus bran mash (Formula 4 or 5). Be sure to mix the bran and poison as a dry mash and scatter in alfalfa fields about 4 o'clock in the afternoon, and around orchard trees or other plants early in the morning.

GUMMOSIS

Not a specific disease. Many different causes. In stone fruit and citrus trees gumming is simply a symptom of distress. May be due to unsuitable soil, poor condition of soil, excess or lack of water, frost or attacks of parasites. Treatment must vary according to cause.

Badly gummed branches may be removed, gummy diseased areas of bark cut out and the wounds treated as in citrus gummosis. Splitting the bark is useless and often harmful.

NEMATODE (EEL WORM)

Do not attempt to grow susceptible crops on infested soil. Keep such areas clean cultivated in summer or in a cereal crop. Grain may be grown in winter. Almost all important crops except cereals and also fruit trees, are attacked by the garden nematode. The beet nematode attacks some other plants, and where it occurs careful rotations should be followed with the total exclusion of beets for many years. Alfalfa is not seriously affected by the common species, but carries it over to future crops. This crop may be safely planted on beet nematode soil. Nematodes are worst on sandy soil.

PHYSIOLOGIAL DISEASES

(Little Leaf, Exanthema, Die-back, Mottled Leaf, Rosette, Bitter Pit, Dry Rot, Blossom-end Rot)

Diseases of a specific nature in which the cause is not known and which seem unlike the usual effects of unfavorable conditions or parasites. Most of these troubles show a relation to soil conditions and occur especially in dry, sandy, gravelly, or hard-pan soils, those very deficient in humus, or under conditions of irregular soil moisture. Trees standing over old barnyards or corrals or where excessive amounts of manure have been applied are also likely to show some of these conditions. The best possibilities of treatment lie along the line of increasing the humus content of the soil by means of green-manure crops and mulches, breaking up all hard-pan and plow-soles, more careful irrigation to insure the maintenance of a proper and uniform moisture condition of the soil down to a depth of several feet, and throughout the season until rains occur, and planting of alfalfa in orchards where plenty of water is available. The soil in areas where these troubles occur should be examined for alkali or other injurious substances. Where any of these diseases are serious and persistent it may be better to grow some other crop than to keep on with one which is seriously affected.

RABBITS

If very abundant, must be fenced out of young orchards and gardens to avoid serious damage. Shooting and poisoning are the principal means of destruction. An application to the trunks of young trees of whitewash containing bitter aloes is sometimes recommended, but this has not shown much value in actual practice. The same may be said in regard to smearing the trunks with blood.

SLUGS AND SNAILS

May be controlled to some extent by sprinkling dry lime dust upon the ground in circles about the plants, upon the leaves of the plants themselves or in any way so that the slugs will come in contact with the lime. A mixture of salt with the lime is sometimes recommended. This, however, is injurious to plants if it comes in contact with them. May also be trapped by laying boards upon the ground near the plants upon which they feed, thus affording a shelter under which they may be found and killed.

SOUR SAP, WINTER INJURY

All the ordinary forms of Sour Sap are due to freezing, alternate warm and cold weather, or other climatic injury in winter. Differences in the effect upon individual trees or orchards are due to differences in condition and susceptibility of the trees, produced mostly by variation in the moisture condition of the soil. Do not force growth late in summer. Irrigate, if possible, about November 1 if no heavy rain has fallen. Whitewash bodies of trees early in November.

SQUIRRELS AND GOPHERS

May be controlled by persistent poisoning, fumigation with carbon bisulfide, trapping and shooting. For poisoning material, the commercial preparations may be used or Formula 33.

Poisoned Fruit.—Strychnine sulfate may be sprinkled over orange halves or watermelon rind, or a solution of 1 ounce of strychnine sulfate dissolved in 1 gallon of boiling water may be used for saturating grain or other material, after allowing the solution to cool.

SUNBURN

Whitewash bodies in fall as well as spring. Shape the trees so as to shade bodies. Cut young trees back well before planting. Shade trunks with shakes or protectors. Do not allow trees to suffer from drouth.

TENT CATERPILLARS

Cut out and burn nests. Colonies collected on the trunk may be killed by spraying with gasoline or oil emulsion. Cut out egg masses at pruning time. Spray with arsenate of lead, 2 pounds to 50 gallons of water.

TREE WOUNDS, PRUNING CUTS

Make a clean, smooth cut, trimming the bark down smoothly to sound tissues around the edges. In the case of branches, make a smooth cut, leaving no projecting stub. Thoroughly cover the wound with Bordeaux paste (Formula 10) and after callus starts to form about the edges, cover with grade D asphaltum or similar material put on in a melted condition. Go over the work occasionally, especially in the fall, and renew the application of asphaltum until wound is entirely healed.

FORMULAS AND DESCRIPTION OF MATERIALS

ARSENICALS

Acid Lead Arsenate (Lead Hydrogen Arsenate, Di-lead Arsenate, often labeled "Standard" or Lead Arsenate).—The acid type of lead arsenate contains more poison per pound than the basic type and is a stronger and quicker-acting poison. It is, however, somewhat susceptible to the action of other chemicals, particularly those of an alkaline nature (such as soaps, lime-sulfur solution, etc.), and is more or less dissolved by them when used as a combination spray. In the moist climates along the coast, or in continuous damp, cloudy weather elsewhere, whether used alone or in combination with other sprays, some of the arsenic is apt to be dissolved and cause serious foliage injury. It is not considered as a safe arsenical for use on stone fruits, beans or other susceptible plants.

Basic Lead Arsenate (Usually labeled "Tri-plumbic" or "Neutral").—The basic type of lead arsenate contains less arsenic per pound than the acid type, and is a weaker and slower-acting poison. It is not decomposed, however, by chemicals of an alkaline nature, such as are usually applied with it as a combination spray, nor by the damp weather of the coast regions. It is considered the only safe arsenical to use on stone fruits, beans or other susceptible plants.

The lead arsenates are usually sold as a paste containing about 50 per cent of water, or as a dry powder. The paste should be thinned somewhat with water and worked into a smooth cream before adding to the spray tank. The powder may be added directly to the tank and mixed by means of the agitator.

For codling moth and most defoliating insects, use:

FORMULA 1

Acid lead arsenate paste	4 to 8 pounds
Water	100 gallons

or

FORMULA 2

Basic lead arsenate paste	5 to 10 pounds
Water	100 gallons

Dry or powdered lead arsenate contains twice as much arsenic as the paste, therefore use only one-half as much in the above formulas.

Zinc Arsenite is a more active and stronger poison than either type of lead arsenate and is useful in controlling the various caterpillars which are troublesome on pears and apples in the early spring, but is very apt to cause injury if the application is made after the time of full bloom.

FORMULA 3

Zinc arsenite powder	3 pounds
Water	100 gallons

White Arsenic (*Arsenic trioxide*) is only sparingly soluble in water, although sufficiently so to prohibit its use on plants as an insecticide. Its use as a stomach poison is therefore limited to the preparation of poison baits, for the control of grasshoppers, armyworms, cutworms, etc., and in some other cases where the insecticide is not to be applied to growing plants.

Poison Bran-mash.—

FORMULA 4

Bran	25 pounds
White arsenic	1 pound
Molasses (cheap blackstrap preferred)	2 quarts

Mix the arsenic and the bran dry, and add the molasses which has been diluted with water. Add enough water and mix thoroughly to make a dry mash which will broadcast easily.

Citrus Bran-mash.—

FORMULA 5

White arsenic	1 pound
Molasses (cheap blackstrap preferred)	2 quarts
Lemons (or oranges)	6 fruits
Water (about)	4 gallons
Bran	25 pounds

Mix the above materials as follows: Stir thoroughly the white arsenic, molasses, and water first. Grind the lemons, including the rinds, in a meat grinder, or chop fine, and add to this liquid. Then slowly pour this over the bran and stir thoroughly until an even mixture is secured.

The amount of water to use in the preparation of these baits will vary according to the coarseness of the bran, or substitutes. A dry mash is preferable to a wet mash because it does not harden under the heat of the sun and remains palatable, while wet mash becomes baked and unattractive.

Substitutes in Poison Baits.—Paris green may be substituted for white arsenic in formulas 4 and 5. Alfalfa meal, shorts, or rice meal, have been successfully used as a substitute for bran in the preparation of the above formulas.

Sodium Arsenite.—This arsenical is readily soluble in water and is one of the most violent of the plant poisons. It is probably the most quick-acting of any of the better known arsenical poisons, and commonly used in the preparation of weed killers, poison fly-papers, cattle dips for the control of ticks, ant syrups, and to some extent in the preparation of poison baits.

Sodium arsenite may be purchased ready-made as a white powder, but it is not always readily obtained at pharmacies, nor is it always dependable in having a uniform amount of arsenic. This chemical can be easily prepared from white arsenic by combining the latter in the presence of water with sal soda, soda-ash, caustic soda, or a good grade of concentrated lye in the following proportions:

Sal soda or washing soda, 2 parts to 1 part of white arsenic.
Soda-ash, 1 part to 1 part of white arsenic.
Caustic soda, 1 part to 2 parts of white arsenic.
Concentrated lye, 1 part to 2 parts of white arsenic.

If sal soda or soda ash is used it is necessary to boil the mixture fifteen or twenty minutes before the arsenic is dissolved. If caustic soda or concentrated lye is used, little or no heat is necessary. In either case, a corrosive chemical is formed known as sodium arsenite.

A soluble arsenical can be made by using one part of caustic soda to four parts of arsenic trioxide. Such a solution, however, has a tendency to form crystals on standing.

Sodium Arsenite.—

FORMULA 6

Sal soda	2 ounces (or 2 pounds)
White arsenic	1 ounce (or 1 pound)
Water (about)	½ pint (or 1 gallon)

Put all the ingredients together in an iron or graniteware kettle (do not use aluminum), of sufficient size to allow for considerable frothing, and boil fifteen or twenty minutes, or until solution is clear.

A modification of Professor C. W. Woodworth's formula which has been successfully used in municipal campaigns against the Argentine ant is as follows:

Argentine Ant Syrup—

FORMULA 7

Sugar	18 pounds
Water	6 quarts

First dissolve the sugar in the water by stirring, or by heating and stirring, then add one ounce of white arsenic which has been previously converted into sodium arsenite, according to the directions given in formula 6, and add enough water to make exactly three gallons. This formula will produce a syrup containing .2 of 1 per cent of arsenic trioxide by weight.

The U. S. Bureau of Entomology recommends a later formula for the preparation of Argentine ant syrup which is said to be superior to any other formula tested by them, "on account of its stability at high temperatures, freedom from crystalization, and continued attractiveness."

Government Argentine Ant Syrup.—

FORMULA 8

Prepare a syrup:

Granulated sugar	15 pounds
Water	7 pints
Tartaric acid (crystalized)	¼ ounce

Boil for 30 minutes. Allow to cool.

Dissolve sodium arsenite (chemically pure)	¾ ounce
In hot water	1 pint

Cool. Add poison solution to syrup and stir well.
Add to poisoned syrup:

Honey	1½ pounds

Mix thoroughly.

COPPER COMPOUNDS

Bordeaux Mixture (Average Strength).—

FORMULA 9

Bluestone	16 pounds
Quicklime	20 pounds
Water	200 gallons

Dissolve the bluestone and slake the lime in separate vats. Thoroughly mix the dissolved bluestone with one-half the water, and the slaked lime with the other half. Run the two mixtures together in a single stream into the spray tank through a fine screen. For convenience the mixing vats may be placed on an elevated platform, and the two parts mixed as they are flowing into the spray tank. The milk of lime should be continuously stirred during the mixing.

A somewhat less satisfactory Bordeaux mixture may be made as follows: Slake the lime and dissolve the bluestone in separate barrels as above. Fill the spray tank half full of water, add the dissolved bluestone, strain in the lime while the agitator is running, add remainder of water, and mix thoroughly.

Bordeaux Paste.—

FORMULA 10

A—Bluestone .... 12 pounds
Water .... 8 gallons
B—Quicklime .... 24 pounds
Water .... 8 gallons

Dissolve the bluestone and slake the lime separately in the amounts of water specified. Then mix together equal quantities of each ingredient, making up only enough for each day's use.

Commercial Bordeaux Mixture.—Several preparations of this sort are on the market in the form of a paste or dry powder to be diluted with water. Objection is sometimes made to these preparations that they will not remain in suspension in water as well as the home-made Bordeaux mixture, but some of them are probably as good or better than the average mixture prepared on the ranch. The commercial preparations are more expensive but more convenient for use, and are especially of interest to the small grower.

Bluestone (Copper Sulfate).—A soluble compound of copper, the raw material for the preparation of most other compounds of copper. This cannot be used on foliage.

For dipping grain use

FORMULA 11

Bluestone .... 1 pound
Water .... 4 gallons
Dip for 3 minutes.

Followed by

Quicklime .... 1 pound
Slaked in water .... 10 gallons

For lemon wash water use

FORMULA 12

Bluestone .... 1½ pounds
Water .... 1000 gallons

SULFUR AND SULFUR COMPOUNDS

Dry Sulfur.—For dusting upon plants for the control of surface mildew, red spider, or other parasites, the fineness of the sulfur is an all-important consideration. Flowers of sulfur, the finest and fluffiest grade of sublimed sulfur, has been heretofore recommended for application as a dust. At present, however, there are upon the market several brands of extremely finely ground sulfurs, which are finer than the best grade of sublimed sulfur and no more expensive. Some of these sulfurs, which have been specially prepared for dusting, are ground to pass a 200-mesh bolting cloth. These are apt to cake or to clog the dusting apparatus. If three parts of sulfur are thoroughly mixed with one part of hydrated lime, kaolin, or other inert powder, these difficulties may be avoided.

Sulfur Pastes or Wettable Sulfurs.—For various reasons it is often desirable to mix sulfur and water and apply to plants as a spray. Sulfur, however, is not easily wetted with water and it is a difficult matter to make a uniform mixture of the two. It has been found that a number of substances, soap, oleic acid, glue, diatomaceous earth, flour, dextrin, etc., when mixed with water and sulfur, have the property of counteracting the natural aversion of sulfur to water without otherwise altering the nature of the sulfur. Certain of these substances have been used in the preparation of commercial sulfur pastes or wettable sulfurs. These commercial pastes, as now manufactured, contain from 45 to 50 per cent of sulfur in a very finely divided condition, the remainder being water and one of the substances mentioned above. The effect of these pastes is that of dry sulfur. The usual strength to use is

FORMULA 13

Commercial sulfur paste .... 8 to 21 pounds
Water .... 100 gallons

Home-made Wettable Sulfur.—A satisfactory wettable sulfur can be easily made at home by the use of glue water as follows:

FORMULA 14

Powdered glue .... ¾ ounce
Hot water .... 1½ gallons
Sulfur (flowers or powdered) .... 5 pounds
Water to make .... 100 gallons

Dissolve the glue in hot water, or soak over night in 1½ gallons of cold water. Add the glue water to the sulfur a little at a time and work up into a smooth paste as free from lumps as possible. Rubbing is better than stirring. Wash this paste into the spray tank through a fine screen, using the remainder of the glue water to wash it through and a stiff brush to break up the remainder of the lumps. Then add plain water to make 100 gallons.

Another formula more expensive than the above is

FORMULA 15

Make a paste of
   Flour .... 4 pounds
   Water .... 4 gallons
Mix this with
   Sulfur (sublimed or powdered) .... 5 pounds
Then add
   Water to make .... 100 gallons

The usual grades of sublimed or powdered sulfur may be wetted in the manner described in Formulas 14 and 15, but if the best results are to be obtained, the finest grade of sulfur obtainable should be used. The sulfurs especially prepared for dusting are highly recommended for this purpose.

Lime-sulfur Solution.—This is the most active form in which sulfur compounds are commonly used in the control of insects or fungi. Its causticity prohibits its use on foliage except that of the more hardy plants, and then in very dilute form. Its largest use is as a dormant spray for the control of certain fungous diseases, scale insects and a variety of other pests of deciduous trees.

Commercial Lime-sulfur Solution.—The growers of the state are being supplied with concentrated commercial lime-sulfur solution of good quality and at reasonable prices. The great bulk of this important pest remedy used in the state is therefore of commercial manufacture, testing between 32° and 34° Baume. It is only necessary to dilute this with water before spraying.

Home-made Lime-sulfur Solution.—

FORMULA 17

Stone Lime	50 pounds
Sulfur (sublimed or powdered)	100 pounds
Water to make	50 gallons

Heat in a cooking barrel or vessel about one-third of the total volume of water required. When the water is hot, add all of the lime, and at once add all the sulfur, which should previously have been made into a thick paste with water. After the lime is slaked, another third of the water should be added, preferably hot, and the cooking should be continued until a clear orange-colored solution is obtained (usually 45 to 60 minutes), when the remainder of the water should be added, either hot or cold as is most convenient. The boiling due to the slaking of the lime thoroughly mixes the ingredients at the start, but subsequent stirring is necessary if the wash is cooked by direct heat in a kettle. After the wash has been prepared it must be allowed to settle and then strained through a fine sieve as it is being run into the spray tank. The resultant product is a concentrated solution of lime-sulfur, which should be diluted about six times with water for a winter spray.

Alkali Sulfides.—Sulfides of soda ("soluble sulfur") are sometimes used in place of lime-sulfur solution and have some advantages over the liquid preparations.

CRUDE PETROLEUM

The use of crude petroleum is almost entirely limited to the winter spraying of deciduous trees when the buds are entirely dormant. It is generally applied from November to February. The crude oil emulsion is especially recommended for black scale (*Saissetia oleae*), European fruit Lecanium (*Lecanium corni*), European or Italian pear scale (*Epidiaspis piricola*), cherry scale (*Lecanium cerasorum*) and other scales infesting deciduous fruit trees. It is practically the only spray treatment which has been effective against the European or Italian pear scale and, to a certain extent, will destroy the winter eggs of many of the aphids, red spider, and some of the defoliating caterpillars.

When crude oil is thoroughly applied it sometimes penetrates the fruit buds to a considerable extent, some of which may be injured and even killed. The great majority of the buds are not injured, however, but appear to be stimulated to a more vigorous growth; producing a character of foliage which withstands attacks of diseases. It is good practice, especially in dry seasons, not to apply crude oil emulsion until there is an indication of the swelling of the buds.

A natural crude petroleum, testing about 23° Baume, is preferred as it contains some of the lighter and more penetrating oils. Heavier crudes than this have given satisfactory results, even those testing 18° and even lower. Residuum oils (the residue of crue petroleum after the lighter portions have been distilled off) can be used if natural crude oil is unobtainable, provided their content of asphaltum is not too high to prevent their emulsification.

Crude Oil Emulsion.—

FORMULA 18

Water	175 gallons
Liquid soap	3 gallons
Natural crude petroleum (21°–24° Baume)	25 gallons

Partly fill the spray tank with water, add the liquid soap, agitate thoroughly for one minute, add crude oil and continue the agitation, while running in the remainder of the water. If liquid soap cannot be obtained, use 20 pounds of fish-oil soap dissolved in 10 gallons of boiling water to which 3 pounds of caustic soda or lye have been added. To kill moss or lichens on fruit trees add 2 pounds of caustic soda or lye to the formula.

During the spraying operation this emulsion should be thoroughly agitated and great care taken to wet all of the twigs. From 8 to 10 gallons should be used on a tree.

PETROLEUM DISTILLATES

Kerosene, of about 40° Baume, has been used to a considerable extent as an insecticide, particularly on citrus trees, applied in the form of an emulsion. The cheaper, unrefined distillates have now largely replaced kerosene as a foliage spray. These are more effective as an insecticide, so that smaller percentages can be used in the emulsions, but coupled with their superior insecticidal properties is their greater toxicity to fruit and foliage. The toxicity of an oil varies with climatic conditions, foliage injury being most certain in dry weather with a temperature of 95° F. or more. Unfortunately, the season when spraying is most effective against scale insects on citrus trees is often during the hottest and driest months. It seems impossible to guarantee immunity from damage with any of the varying distillates obtainable, irrespective of climatic conditions or the condition of the trees.

Less injury to citrus fruit and foliage occurs in the coast regions where distillate emulsions have been used, but in the interior sections the use of this insecticide is very hazardous.'

Spraying with distillates, or with any other material, is not recommended as a substitute for fumigation in commercial citrus orchards, except in case of young orchards, trees about dooryards, or where fumigation may not be convenient, or infestation may be light or limited to occasional trees. In such cases, Formula 19 is considered the most satisfactory.

Kerosene emulsion is the safest of the petroleum-distillate sprays, although the most expensive. The "W.W." or "Water White" is a trade name of a low-grade kerosene and is safer than the usual grade of material sold as "distillate." The highly refined "case goods" kerosene has been found to cause the least amount of injury of any of the petroleum derivatives, but its cost would prohibit its use except on a small scale. If much of the kerosene emulsion is allowed to run down the trunks of young trees, injury is likely to occur just beneath the surface of the ground.

The following formula is intended for use on citrus trees:

Kerosene Emulsion.—

FORMULA 19

Kerosene	15 gallons
Liquid soap	¾ gallon
(Or hard soap	4 pounds)
Water	200 gallons

If liquid soap is available, it is preferable to hard soap, since no heating is required. Hard soap, preferably fish-oil, is cut in thin slices and dissolved in hot water. The soap is placed directly in the spray tank with 10 or 15 gallons of water or more (the exact amount is not important), and then the engine is started. The oil is now added slowly, and the materials are emulsified by being run through the pump under pressure. After a few minutes the rest of the water may be added, and the spray is ready to apply to the trees.

Certain "tree" distillates, testing 31° to 32° Baume, said to be selected and partially refined, have lately displaced to a considerable extent the heavier distillates of 27° to 28° for use on citrus trees.

"Tree" Distillate Emulsion.—

FORMULA 20

Tree distillate (31°–32° Baume)	4 gallons
Liquid soap	¾ gallon
(Or hard soap	5 pounds)
Water	200 gallons

These materials are emulsified in the same manner as explained for the kerosene emulsion, Formula 19. If the distillate is used without soap, the following is the formula:

Straight "Tree" Distillate.—

FORMULA 21

Tree distillate (31°–32° Baume)	4 to 6 gallons
Caustic soda (95 per cent)	7 pounds
Water	200 gallons

In the case of the straight distillate, the oil is kept in suspension in the water by agtitation and forms an unstable mechanical emulsion, which separates quickly in standing. In using this, it is necessary to have the spray outfit equipped with a powerful and efficient agitator, which must be kept going continuously during the spraying operations.

The use of petroleum-distillate sprays against black scale on olive trees is now being recognized as a profitable practice. For this purpose the heavier distillates of 28° to 30° Baume are used since olive foliage is very resistant to spray injury and also because the spray can be applied through the winter months when low temperatures and high humidities are the rule.

Distillates of this density are also much used as a dormant spray on deciduous trees, although crude oil sprays are replacing more and more the distillates for this purpose.

*Heavy Distillate Emulsions.—*For use on olives, the following mechanical emulsion is recommended:

FORMULA 22

Distillate (28° Baume) .... 7 gallons
Caustic soda (95%) .... 5 to 7 pounds
Water ....to make 200 gallons

First dissolve the caustic soda in a small amount of water and add to the water in the spray tank; begin the agitation and slowly add the distillate, continuing the agitation during the application. This spray will also remove lichens or moss from the trees.

By reducing the amount of crude oil from 25 gallons to 15 gallons in Formula 18, the crude oil emulsion may be used on olive trees for the control of black scale.

For use on deciduous trees the following is recommended:

FORMULA 23

Distillate (27°–28° Baume) .... 20 gallons
Fish-oil soap .... 30 pounds
Water to make .... 12 gallons

Dissolve the fish-oil soap in water, heating it to the boiling point, add the distillate and agitate thoroughly while the solution is hot. For use add 20 gallons of water to each gallon of the above mixture.

Commercial Prepared Emulsions and Miscible Oils.—Many growers realize the difficulty in securing proper materials for home-made emulsions and the variability of the home-made mixtures even under the best conditions. They prefer to buy manufactured products, especially when only small quantities are needed. The commercial emulsions and miscible oils are no more effective than a good home-made preparation and are only of interest as a matter of convenience. These preparations are on the market in great variety, many of which are sold under trade names. Practically all grades of petroleum distillates, as well as crude petroleum, are obtainable in a form ready to be used, after simple dilution with water. If these ready-made preparations are to be used, it is especially important to purchase only from reliable and well-known manufacturers or dealers. The commercial products in general are satisfactory for use for the purposes indicated in the above formulas.

The following is recommended for the control of thrips:

Distillate Emulsion and Tobacco Extract.—

FORMULA 24

Water .... 12 gallons
Fish-oil soap .... 30 pounds
Distillate (32°–34° Baume) .... 20 gallons

The above emulsion is prepared in the ordinary way as a stock solution. For use in the orchard dilute 1 to 20 parts of water. To every 200 gallons of this diluted spray add 1 pint of tobacco extract containing 40 per cent nicotine, or about 3½ gallons of tobacco extract containing 2¾ per cent nicotine.

The Rosin Wash is chiefly used for young and tender nursery stock, because it does not cause the injury often following the application of petroleum distillates. The preparation is

FORMULA 25

Rosin .... 10 pounds
Caustic soda (76% to 95%) .... 3 pounds
Fish oil .... 1½ pounds
Water to make .... 50 gallons

To a gallon of hot water in an iron kettle add the fish oil and the rosin and heat until the latter is softened; after first dissolving the caustic soda in a small quantity of water add it and stir the mixture thoroughly, after which pour in enough water to make 50 gallons of spray material.

Crude Carbolic Acid Emulsion.—

FORMULA 26

Fish-oil soap .... 40 pounds
Crude carbolic acid .... 5 gallons
Water to make .... 40 gallons

Dissolve the soap in hot water completely, add the carbolic acid and heat to the boiling point for twenty minutes (reserve some water to add in case the mixture begins to boil over). For use add 20 gallons of water to every gallon of the above stock solution. The emulsion needs little or no agitation.

TOBACCO PREPARATIONS

Concentrated commercial preparations of tobacco have almost entirely superseded the home-made tobacco infusions on account of their greater convenience and uniformity. A material containing 40 per cent nicotine in the form of nicotine sulfate is recommended for the preparation of contact insecticides containing nicotine. The usual formula is

FORMULA 27

Tobacco extract (nicotine sulfate 40%) ...... 1 pint
Fish-oil soap .... 4 to 5 pounds
Water .... 100 to 150 gallons

For small quantities use 1 teaspoonful to 1 gallon of water.

Tobacco Dust.—Finely-ground tobacco dust finds some use as insecticide, particularly in the control of aphids. Fifty per cent of kaolin or hydrated lime is sometimes mixed with it as a diluent.

MISCELLANEOUS

Corrosive Sublimate (Bichloride of Mercury).—This is a very poisonous substance and is one of the most powerful of germicides; it is employed to some extent in plant-disease treatment. The usual strength is

FORMULA 28

Corrosive sublimate	1 ounce
Water	8 gallons,

or 1 part to 1000.

Tablets to make this strength when added to 1 pint of water may be obtained at drug stores. Distilled or rain water should be used; the solution must not be kept in a metal container.

Whitewash.—

FORMULA 29

(Ordinary Formula)

Water	2 gallons
Quick-lime	10 pounds

Add more water after slaking to bring the wash to the desired consistency.

A more durable white-wash:

FORMULA 30

Quick-lime	5 pounds
Salt	½ pound
Sulfur	¼ pound

Slake the lime slowly with water and add the salt and sulfur while it is boiling. Add enough more water to make a good wash. This is good for white-washing the bodies of trees in the fall.

Government Whitewash.—

FORMULA 31

Quick-lime	40 pounds
Salt	15 pounds
Rice Flour	3 pounds
Spanish whiting	½ pound
Glue	1 pound
Water	5 gallons

This is complicated and expensive, and some of the ingredients are often difficult to obtain.

Grafting Wax.—

Many different combinations are used for this purpose, most of them being various combinations of beeswax and rosin. The following formula is a good one:

FORMULA 32

Rosin	4 pounds
Beeswax	1 pound
Linseed oil	1 pint

The ingredients are all melted and mixed together in a kettle. In hot weather use more rosin.

Some use one pound of tallow as a substitute for the linseed oil. One ounce of lamp-black or one pint of flour is sometimes added. Asphaltum is used to some extent as a substitute for rosin and beeswax and, in fact, straight asphaltum is used successfully in some cases for grafting wax.

Carbon Bisulfide. is a liquid which evaporates quickly when exposed to the air, forming a heavy and inflammable vapor of a great penetrating power. In using the material for fumigation, it is essential that it be placed at the top of the room in a shallow container in order that the heavy vapors as they are driven off will thoroughly diffuse with the air contained in the space to be fumigated. The proper amount to use depends upon the type of room being fumigated and ranges up to about 30 pounds to 1000 cubic feet in ordinary rooms where the walls and floor have not been made especially tight. The best results are obtained by doing this work when the temperature is above 75° F.

Carbon bisulfide is one of the best agents for destroying ground squirrels that have failed to take poisoned grain, or having once survived the poison refuse to take it again. It is recommended for use against ground squirrels during the winter season when the ground is wet.

The two best methods of applying carbon bisulfide are by the use of the "waste-ball" method and of the "destructor." The common waste-ball method is to pour a tablespoonful of carbon bisulfide on a piece of cotton waste, corn cob, horse manure, or other absorptive material, which should then be thrown as far down the hole as possible and the opening immediately closed with earth. Explosion of the gas in connection with the waste-ball method is recommended where the ground is damp and there is no danger from fire. The explosion of the gas forms new compounds which are poisonous and may diffuse somewhat more rapidly than the vapors of the material. A method which is said to be equally effective to exploding the gas as above is by the use of the "destructor," which pumps the vaporized carbon bisulfide into the burrow.

Poisoned Barley.—Following is the latest government formula for preparing poisoned barley for California ground squirrels:

FORMULA 33

Barley (clean grain)	16 quarts
Strychnine (powdered alkaloid)	1 ounce
Bicarbonate of soda (baking soda)	1 ounce
Thin starch paste	¾ pint
Heavy corn sirup	¼ pint
Glycerin	1 tablespoonful
Saccharin	1/10 ounce

Mix thoroughly 1 ounce of powdered strychnine and 1 ounce of common baking soda. Sift this into ¾ pint of thin, hot starch paste and stir to a smooth, creamy mass. (The starch paste is made by dissolving 1 heaping tablespoonful of dry gloss starch in a little cold water. which is then added to ¾ pint of boiling water; boil and stir constantly until a clear thin paste is formed.) Add ¼ pint of heavy corn sirup and 1 tablespoonful of glycerin and stir thoroughly. Add 1/10 ounce of saccharin and stir thoroughly. Pour this mixture over 16 quarts of clean barley and mix well so that each grain is coated.

Caution: All containers of poison and all utensils used in the preparation of poisons should be kept *plainly labeled* and *out of reach* of children, irresponsible persons and live stock.

INDEX

Alfalfa 1, Nematode 17, in orchards 17, meal 21.

Alkali 17, sulfides 25.

Angoumois moth 9.

Anthracnose bean 4.

Ants 15, Argentine 15, sirups 21 and 22.

Aphids 15, oil emulsion for 26, tobacco dust for 31.

Aphis, green and rosy apple 2; green, rosy, wooly 3; bean 4, cabbage 5, celery 6, cherry 6, chrysanthemum 6, citrus 7, cucumber 8, dahlia 8, melon 9, pea 10, green on pear root 12, sorghum 13, strawberry 14, walnut 14.

Apoplexy grape 9.

Armillaria, almond 1, apricot 3, citrus 6, olive 10, peach 11, plum and prune 12, walnut 14, **18.**

Armyworms, alfalfa 1, beet 5, corn 7, grain 8, grape 9, onion 10, pea 10, potato 13, sorghum 13, tomato 14, **16.**

Arsenate of lead, cherry 6, grape 9, red humped caterpillar 12, cankerworms 15, **19.**

Arsenic, white 20, trioxide 20.

Arsenicals 19.

Asphaltum, peach 11, crown gall 16, wounds, pruning cuts 19, residium oils 26, grafting wax 32.

Bark beetles, apricot 4, olive 10.

Barley 4, straw 7, 8, sensitive to bluestone 8, poisoned 23.

Beans 4, Black eyes 5, arsenicals on 19.

Beeswax 32.

Beetles, asparagus 4, cucumber 8.

Bichloride of Mercury 31.

Bitter pit, olive 10, 17.

Black heart, apricot 3.

Black knot, grape 9.

Black leaf, pear 11.

Black rot, sweet potato 14.

Black scale 3, olive 10, plum and prune 12, 26, olives 28.

Blight, apple 2, beet 5, celery 6, pea 10, peach 11, pear 11, late tomato 17, walnut 14.

Blister mite, pear 11.

Blossom blight, apricot 3.

Blossom end rot, tomato 14, **17.**

Blue mold, citrus 7.

Bluestone 7, smut 8 and 23, lemon wash 23.

Bordeaux, apple scab 2, celery blight 6, paste 7, peach 11, pear scab 11, flea beetles on potatoes 13, paste 14 and 16, paste 19 and **27.**

Bordeaux mixture, apple scab 2, bush fruit 5, citrus 7, melons 9, pear 12, strawberry 14, tomato 14, commercial 23.

Borers, shot hole 4, bush fruits 5, currant and gooseberry 8, deciduous nursery stock 10, California peach 11.

Brown rot, apricot 3, citrus 7, peach 11, plum and prune 12.

Bud blight, apricot 3.

Burlap covers 16.

California grape root worm 9.

California peach borer, almond 1, apricot 4, cherry 6, peach 11, plum and prune 12.

Cane blight, bush fruits 5.

Canker worm, apple 2, cherry 6, peach 11, plum and prune 12, 15.

Carbolic acid, emulsion 5 and 30.

Carbon bisulfide, bean weavil 4, grain 9, potato tuber moth 13, squirrels and gophers 18, **32.**

Caterpillar, see tent and red humped caterpillars, alfalfa 1, celery 6, hollyhock 9, red humped apple 15, yellow necked apple 15, pears and apples 20, tent 18, 26.

Caustic soda 21, 26, 28, 30.

Centipede 4.

Citrus 6, nursery stock 9, frost 16, bran mash, 16 and **20,** 27.

Codling moth, apple 2, 3; pear 11 and 12, 19.

Combined spraying, almond 1, apple 2, apricot 4, peach 11, pear 12, plum and prune 12, rose 13.

Copper compounds 22.

Copper sulfate 23.

Corrosive sublimate, pear blight 11, potato 13, crown gall 6, **31.**

Cottony rot, citrus 7.

Crown borer, strawberry 14.

Crown gall, alfalfa 1, almond 1, apricot 3, grape 9, 10, peach 11, plum 12, walnut 14, 15.

Crude oil emulsion 12, 26.

Crude petroleum 26.

Curly top, beet 5.

Cutworms, alfalfa 1, beet 5, corn 7, grain 9, grape 9, onion 10, pea 10, potato 13, sorghum and Sudan grass 13, tomato 14, **16.**

Damping off, citrus 6, tomato 14, 16.

Deciduous trees, sprays for 28.

"Destructor" for carbon bisulfide 32.

Dextrin 24.

Diabrotica, dahlia 8.

Diatomaceous earth 4.

Dieback, cherry 6, olive 10, 17.

Distillates 27, tree 28, heavy 28, 29.

Dry rot, olive 10, potato 12, 17.

Ear mold, corn 7.

Ear worm, corn 7.

Eel worm 17.

Emulsion, crude oil 26, kerosene 27, "tree" distillate 28, distillate and tobacco extract 29.

Emulsions, heavy distillate 28, commercial 29.

European fruit lecanium 26.

Eutettix tenella 5.

Exanthema, olive 10, 17.

Fish oil soap 27, 29, 30.

Flat headed apple tree borer, apple 2, plum 12.

CPSIA information can be obtained
at www.ICGtesting.com
Printed in the USA
BVHW051024160423
662436BV00009B/201